CONSERVATION CHAMPIONS!

CONSERVING ELECTRICITY

SADIE SILVA

Enslow PUBLISHING

Please visit our website, www.enslow.com.
For a free color catalog of all our high-quality books, call toll free 1-800-398-2504 or fax 1-877-980-4454.

Library of Congress Cataloging-in-Publication Data
Names: Silva, Sadie author
Title: Conserving electricity / Sadie Silva.
Description: Buffalo, NY : Enslow Publishing, [2027] | Series: Conservation Champions! | Includes index.
Identifiers: LCCN 2025053590 (print) | LCCN 2025053591 (ebook) | ISBN 9781978547711 library binding | ISBN 9781978547704 paperback | ISBN 9781978547728 ebook
Subjects: LCSH: Electric power–Conservation
Classification: LCC TJ163.35 .S535 2027 (print) | LCC TJ163.35 (ebook) | DDC 333.791/6–dc23/eng/20260206
LC record available at https://lccn.loc.gov/2025053590
LC ebook record available at https://lccn.loc.gov/2025053591

Published in 2027 by
Enslow Publishing
2544 Clinton Street
Buffalo, NY 14224

Designer: Tanya Dellaccio Keeney
Editor: Therese Shea

Photo credits: Cover (house) Lydia Saccio/Shutterstock.com; cover (girl) Mariana Serdynska/Shutterstock.com; p. 5 (top) Thomas Holt/Shutterstock.com; p. 5 (bottom) fizkes/Shutterstock.com; p. 7 Martin Bergsma/Shutterstock.com; p. 9 Bilanol/Shutterstock.com; p. 11 (bottom) CL Shebley/Shutterstock.com; p. 11 (top) Maksim Safaniuk/Shutterstock.com; p. 13 PeopleImages/Shutterstock.com; p. 15 Only_NewPhoto/Shutterstock.com; p. 17 (top) wavebreakmedia/Shutterstock.com; p. 17 (bottom) New Africa/Shutterstock.com; p. 19 DC Studio/Shutterstock.com; p. 21 cunaplus/Shutterstock.com.

Printed in the United States of America

CPSIA compliance information: Batch #CSENS27: For further information contact Enslow Publishing, at 1-800-398-2504.

CONTENTS

BOLDFACE WORDS APPEAR IN THE GLOSSARY.

ELECTRIC ENERGY

Many machines in your home need power, or energy, to work. TVs, lamps, toasters, and computers need power. So do washing machines, dryers, and fans. Nearly all homes in the United States use electricity as a **source** of power.

SMARTPHONES AND TABLETS USE ELECTRICITY. THEIR **BATTERIES** NEED ELECTRICITY TO CHARGE UP.

WHAT IS ELECTRICITY?

Electricity is the flow of tiny **particles** called **electrons**. Electricity likely comes to your home through power lines. These lines connect to a power station, where electricity is produced, or made. We need electricity, but we should also conserve it. That means not waste it.

DIFFERENT POWER STATIONS USE DIFFERENT ENERGY SOURCES TO PRODUCE ELECTRICITY.

DON'T WASTE IT

Why should you only use the electricity you need? One important reason is that many power stations burn **fossil fuels** to **produce** electricity. When fossil fuels burn, they **pollute** the air. People and animals can get sick from this pollution.

FOSSIL FUELS, SUCH AS COAL, ALSO GIVE OFF GASES THAT TRAP THE SUN'S HEAT. TOO MUCH HEAT HARMS THE **ENVIRONMENT**.

Some ways of producing electricity pollute less. Still, more than half of all electricity in the United States is produced using fossil fuels. Most is made by burning **natural gas**. It is important to conserve electricity until more electrical power comes from better sources.

WIND POWER AND SOLAR, OR SUNLIGHT, POWER ARE TWO SOURCES OF ELECTRICITY THAT DON'T ADD TO AIR POLLUTION.

WIND TURBINES

SOLAR PANELS

WHAT CAN YOU DO AT HOME?

You can be a conservation hero! You can conserve electricity in your own home. For example, you can turn off lights when you leave a room. You can unplug electronic **devices** when you don't need them. You can unplug your devices' chargers too. That saves electricity.

GOING OUTSIDE TO PLAY INSTEAD OF USING ELECTRONIC DEVICES IS A FUN WAY TO CONSERVE ELECTRICITY.

Many homes have electric water heaters. In those homes, using less hot water means conserving electricity *and* water. Showers use less water than baths. So, taking a short shower instead of a bath is a smart way to conserve.

AIR CONDITIONERS USE A LOT OF ELECTRICITY. CLOSING WINDOWS AND SHADES IN SUMMER CAN KEEP A HOME COOL.

WHAT CAN YOU DO AT SCHOOL?

Schools are another great place to conserve electricity. You can ask your teachers to turn off unneeded lights. Computers can be shut down at the end of the day. Reusing and recycling things such as paper and plastic also conserves electricity.

IT TAKES ELECTRICITY TO MAKE PAPER AND PLASTIC. SO, REUSING AND RECYCLING THESE THINGS ARE WAYS TO CONSERVE.

TALK ABOUT IT

Just noticing what uses electricity can help you conserve more. Telling friends what you are doing can help too. They can start conserving electricity and help you think of new ways to conserve. A whole team of conservation champions can really make a difference!

PLANTING TREES CAN HELP CONSERVE ELECTRICITY TOO. TREES PROVIDE SHADE THAT MAKE BUILDINGS COOLER IN SUMMER. TREES HELP THE ENVIRONMENT IN MANY WAYS!

ACTIVITY

- Walk around your home. Make a list of all the devices that use electricity.
- Put a mark next to everything that could be turned off or even unplugged sometimes.
- Show an adult in your home your list. Ask them if they agree.
- Make a family plan to conserve electricity.
- What role will you have?

GLOSSARY

air conditioner: A device that controls the amount of water in the air as well as air temperature.

battery: A device that turns chemical energy into electricity.

device: A tool or machine made to perform a task.

electron: A particle found in atoms that acts as a carrier of electricity in solids.

environment: The natural world in which a plant or animal lives.

fossil fuel: Matter formed over millions of years from plant and animal remains that is burned for power.

natural gas: A fossil fuel mostly made of the gas methane that formed deep underground over millions of years from plants and animals.

particle: A very small piece of something.

pollute: To become unsafe for people or animals.

source: A usable supply of something.

FOR MORE INFORMATION

BOOKS

Forde, Cindy. *Bright New World: Building a Better Planet*. London, England: Welbeck Editions, 2022.

Jackson, Tom. *Matter & Energy.* Minneapolis, MN: Bearport Publishing Company, 2025.

WEBSITES

Electricity for Kids
www.ducksters.com/science/electricity_101.php
Read more about electricity and how it's produced.

Energy Facts for Kids
www.constellation.com/energy-101/energy-choice/energy-activities-for-kids.html
Learn more about different types of energy and ways to conserve.

Publisher's note to educators and parents: Our editors have carefully reviewed these websites to ensure that they are suitable for students. Many websites change frequently, however, and we cannot guarantee that a site's future contents will continue to meet our high standards of quality and educational value. Be advised that students should be closely supervised whenever they access the internet.

INDEX